03565450 9

In the same series

TELEPHONE BOXES by Gavin Stamp

SHOP FRONTS by Alan Powers

TROUGHS AND DRINKING-FOUNTAINS by Philip Davies

HERALDRY by John Martin Robinson

PILLAR BOXES by Jonathan Glancey

John Sambrook

FANLIGHTS

Chatto & Windus
LONDON

Published in 1989 by
Chatto & Windus Ltd
30 Bedford Square
London WC1B 3SG

A CIP catalogue record for this book
is available from the British Library.

ISBN 0 7011 3506 9

Cover: An Irish fanlight, possibly in County Kilkenny, photographed by the author's mother in 1969. A copy of the book will be sent to the first person to identify it.

Title page. Merron Square, Dublin.

Photoset and printed in Great Britain by
Redwood Burn Limited, Trowbridge, Wiltshire

CONTENTS

Two designs by Joseph Bottomley *c.*1793 (*Soane Museum*)

For my onetime colleague
GRAHAM FINCH
because he badgered me into making a fanlight

PREFACE

The word fanlight is commonly misapplied to that inelegant, if practical, part of a casement window more properly called a top-hung quarter-light – and even more commonly misapplied, alas, by being built into the top of a hardwood door. This book is about the decorative lights over the doors of Georgian houses; they were truly Georgian, because they originated soon after the accession of George I, and disappeared early in the reign of Victoria.

When the London Building Act of 1774 banished projecting wooden doorcases, the decoration of ordinary terrace houses came to rely on fanlights and iron railings, and for the next sixty years fanlights contributed the only touch of whimsy to the rather prim face of the late Georgian street. At the beginning of the last war, on the personal and misguided initiative of Lord Beaverbrook, every iron railing that could be taken away without exposing the public to danger was removed for scrap. Without the embellishment of railings, and with – more often than not – plain glass in the fanlights, some streets can seem very dull indeed. Now that everyone travels around by car such things are, I suppose, less noticeable than they were, but to a traveller on foot the innumerable variations contrived from a few glazing bars within a semicircle can be a source of much pleasure and amusement.

It is a great pity that so little has been done since the war to replace these missing railings and fanlights. Even today, when there is unprecedented interest in the preservation and restoration of historic buildings, too little attention is paid to encouraging trainees in the necessary trades. If buildings of architectural value are to be restored successfully, and not debased by feeble, make-do imitations (as is only too evident in the case of fanlights), defunct crafts which were once widely practised need to be researched and sponsored by an appropriate government agency.

When working for the GLC Historic Buildings Division, I came to be responsible for a small store of objects salvaged from demolished houses in the days before architectural salvage became commercial. The collection included a number of damaged fanlights, and it was through contact with this material that the different, ingenious methods used by Georgian fanlight makers became clear to me. Even so, I am very conscious, in writing this account, that there is much yet to be discovered, particularly about developments during the middle years of the eighteenth century when metal glazing bars were first introduced.

The investigation (and, indeed, photography) of fanlights in the field is hampered by the many layers of paint that have accumulated over a century or two; nor can one reasonably expect owners to let it be chipped away in the interests of curiosity. Nevertheless I have been received with great courtesy by all those whose fanlights I have felt it necessary to examine more closely, or check with a magnet; to the far greater number whose fanlights I have simply photographed without asking I offer my apologies. I am most grateful to many of my former colleagues for their advice and assistance; Neil Burton, John Greenacombe, Frank Kelsall and Robert Thorne have been especially helpful, and Robin Wyatt has generously allowed me to copy some of the plates from his collection of early pattern books. I am indebted to Hentie Louw, of the University of Newcastle upon Tyne, for his invaluable work on patents relating to eighteenth-century metal windows. I would also like to thank the National Monuments Record, Sir John Soane's Museum and the Victoria and Albert Museum for assistance with the illustrations. Finally I must acknowledge the contribution made by Isabelle, my wife, who has supported me beyond the call of marital duty.

FANLIGHTS

A fanlight is a window set above an entrance with the object of letting light into the hall. Simple 'over-door lights' probably orginated in the smaller terrace houses of the late seventeenth century, where halls were too narrow to accommodate normal windows. If they had remained at this basic level there would be little more to say, but early in the eighteenth century they began to be treated as a decorative feature. The term 'fan light' seems to have been coined around 1770 to describe the sort of semi-circular overdoor which had a number of panes radiating, like a fan, from a central floret, but within a few years the term was being used to describe all types. Towards the end of the period in which they were fashionable, a fanlight could be defined as a semi-circular or rectangular window, above and enclosed within the same void as a door, in which several panes of glass, *fixed into glazing bars with putty*, were arranged for decorative effect.

The definition includes a reference to the use of putty, so making a clear distinction between fanlights, on the one hand, and, on the other, the decorative wrought-iron grilles used on the Continent, and the older craft of making leaded lights and stained glass. These traditions met, even in the eighteenth century, but they did not become hopelessly confused until the present century.

Window glass

To understand why decorative fanlights came into being we must look, firstly, to the manufacture of glass, because, like many other architectural features, fanlights were a creative response to both the limitations and the possibilities of the materials available.

There were two sorts of window glass available in the eighteenth century: 'broad' glass, which, as its name implies, came in relatively large sheets, and 'crown' glass. Broad glass was blown as an elongated bubble, as much as eighteen inches in diameter and sixty inches long, which was cut up, whilst red hot, and opened out onto a heated metal plate. Contact with the plate, and with the wooden bats used to flatten it, resulted in a finish of such indifferent quality that broad glass was not favoured for windows other than garrets and servants' rooms. Visitors to Erddig, the National Trust house in Clwyd, may judge for themselves the qualities of the broad glass used in the principal rooms facing the garden – although it must be admitted that the present brownish tint, resulting from the use of manganese as a decolouring agent, would not have been apparent when the glass was new.

The best window glass was produced by the crown method. In this a globe was blown and attached, by a nodule of molten glass, to a plain iron rod, or 'punty', at a point directly opposite the entry of the blowpipe, which was then broken free. The incomplete globe, attached to the punty, was then handed to a skilled glassmaker who kept it spinning rapidly in a reheating furnace until the glass had softened and become sufficiently mobile, when it suddenly flew out into a disc about four feet in diameter. The moment this happened it was removed from the furnace, still spinning until it was cool enough to be cut from the punty. The disc, or 'table', of glass was then removed to an annealing furnace (p. 3, *opposite*).

Although tables of crown glass involved much wastage in the cutting, because of their circular shape and the more-or-less useless and uncuttable central bullion, where the punty had been attached, and were subject to slight concentric ripples, they were unmatched for clarity and surface lustre. Crown glass was universally specified for the better sorts of work until well into the nineteenth century, and the preference for it is indicated by the prices given for Crown and Newcastle (the main centre for making broad glass) glass in *The Universal Pocket Companion* of 1741:

A table of crown glass attached to the punty (*Sydney W. Newberry*).

Crown glass in sashes,	nett per foot super	11*d*
Newcastle glass in sashes,	nett per foot super	6*d*

These price differences are even more marked when a deduction of about 2*d* per foot is made for the glazier's labour, making crown glass more than twice as expensive as broad glass.

Whilst the English were filling their elegant sashes with crown glass, Continental glassmakers had perfected the cylinder method and, in 1832, Lucas Chance began making improved 'sheet' glass at his factory at Stourbridge. By cutting the cylinders cold, with a diamond, and opening them out, in a special flattening furnace, onto a heated bed of plate glass the traditional defects of broad glass were avoided. The advantages of good quality, large sheets and lack of waste began to outweigh the superior lustre of crown glass, and finally displaced it when, in 1839, Sir James Chance invented a means of polishing sheet glass without rendering it completely flat. Today a few small tables are still spun by hand but only for the decorative effect of the central bullions – the very bits that a Georgian glazier would have discarded, or sold off cheap to those who could afford nothing better.

Until 1839, then, the maximum size of a pane of the best quality glass was limited by the size and shape of a table of crown glass and, to accommodate it, every window, in all but the poorest classes of building, had to be sub-divided by glazing bars. Given this, and the focal position that an overdoor occupies, it was inevitable that, sooner or later, a decorative virtue would be made of this practical necessity.

Origins

There are few documentary sources for investigating the development of the fanlight. No detailed building accounts have survived for the sorts of houses in which fanlights were likely to have originated, and the large body of surviving architectural drawings, most of which relate to major buildings, are unhelpful because contemporary architects employed a convention, derived from Palladio, in which window and door openings were represented as dark voids. The remaining sources are the craftsmen's exemplars and pattern books and the buildings themselves, and of these the latter must be treated with caution because fashionable improvement and modernisation were just as prevalent in the eighteenth century as they are today. It is likely that some isolated examples of decorative door lights, without relevance to later developments, appeared at a very early date. For example, the White Hart Inn at Scole – halfway between Ipswich and Norwich – is a large coaching inn built in 1655, which has a rectangular overdoor glazed in a lozenge-and-oval pattern. Whether this rather Elizabethan motif is contemporary remains to be proved, but it occurs again on the strings of the staircase.

The most plausible starting-point for a semi-circular fanlight is the standard method of glazing the upper part of a round-headed sash window, where the semi-circle is divided by a pair of glazing bars radiating from a centre or a smaller semi-circle. This is probably the most common design of all and should, perhaps, be termed an 'unconscious fanlight' because it has no decorative intentions. However, in practice, the earliest known fanlights are not of this type but are formed, either wholly or in part, of fretted openings in a solid board. Fretted and spoked fanlights occur in a number of London houses dating from the 1720s, notably at No. 36 Bedford Row, in a late seventeenth-century house, and several in Great James Street, all under hooded and bracketed doorcases (p. 32), and No. 1 Montpelier Row, Twickenham, dated 1720. There is no documentary evidence that these fanlights are contemporary with the houses. Such corroboration is, however, available for the fanlight at Marble Hill House, Twickenham, only a stone's throw from Montpelier Row.[1] Built in 1724–7 for a former mistress of the Prince of Wales, Marble Hill was, nominally, the work of Roger Morris although the real authors of the design were Colen Campbell and his patron Henry, Lord Herbert, later ninth Earl of Pembroke. When Campbell published the third volume of *Vitruvius Britannicus* (1725) he included an elevation of 'A house in Twittenham ...' showing Marble Hill almost as built; most unusually, he sketched in the six-panel door and fanlight, unlike the windows which are represented by the customary dark voids (p. 32 a). Such a coincidence of built and documentary evidence is as valuable as it is rare, and the fact that Campbell troubled to detail it suggests that the fanlight was not commonplace. Similar fanlights occur at Lydiard Park, Wiltshire and Moreton House, Dorset, both of the 1740s and attributed to Roger Morris, and in a house of 1723 in Stamford where it may, or may not, be contemporary.

Pattern books

From about 1725 the rules of Palladian classicism were brought to the attention of lesser architects and craftsmen by a stream of exemplars and pattern books. These ought to be a profitable source of information on the proper decorations for overdoors, and the fact that most of the earlier publications make no mention of them tends to confirm the impression that they were not generally considered as a decorative feature until nearer the middle of the century. If most fanlights were of the 'unconscious' type there would, in any case, be no need to detail what was common joiners' practice. The doorcases illustrated by James Gibbs in his *Book of Architecture*, published in 1728, are clearly not intended to accommodate fanlights. Several doorcases of this type occur in Montpelier Row, and it is instructive to see one or two remaining without fanlights alongside similar examples that have acquired them later (p. 7, *opposite*).

Batty Langley, the carpenter-turned-architect and prolific publisher, whose enthusiasm for the whimsical prompted him to invent, in 1740, a series of Gothick Orders,[2] merely gives some designs for gothick windows which could easily be, and doubtless were, adapted for use as overdoors. In *The Builder's Jewel*, published in 1741, there are excellent engravings showing the proper proportions for classical doorcases but, in common with his fellow architects, he leaves doors and glazing blank (p. 8). He did not, however, carry the impost mouldings across the opening to form a transom, dividing door from

a

b

Doorcases of *c.*1720 in Montpelier Row, Twickenham; *a* as built and, *b* with later fanlight.

a

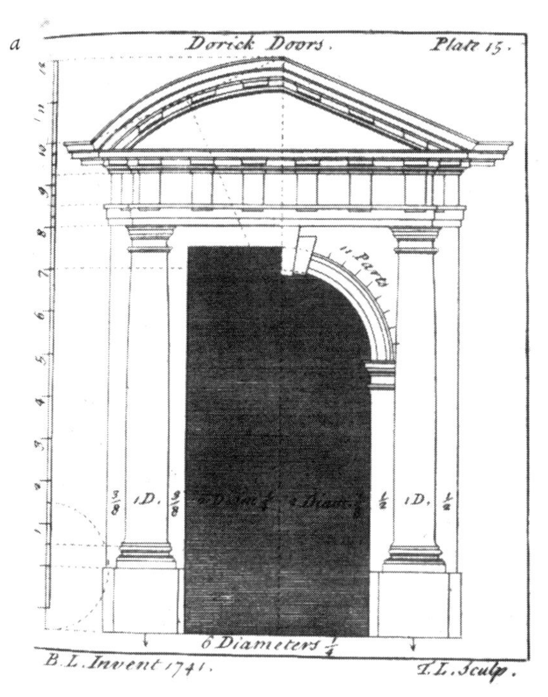

Dorick Doors.
Plate 15.
11 Parts
1 D.
1 D.
6 Diameters 1/4
B. L. Invent 1741.
T. L. Sculp.

b

PROFESSIONAL ASSOCIATION OF TEACHERS

c

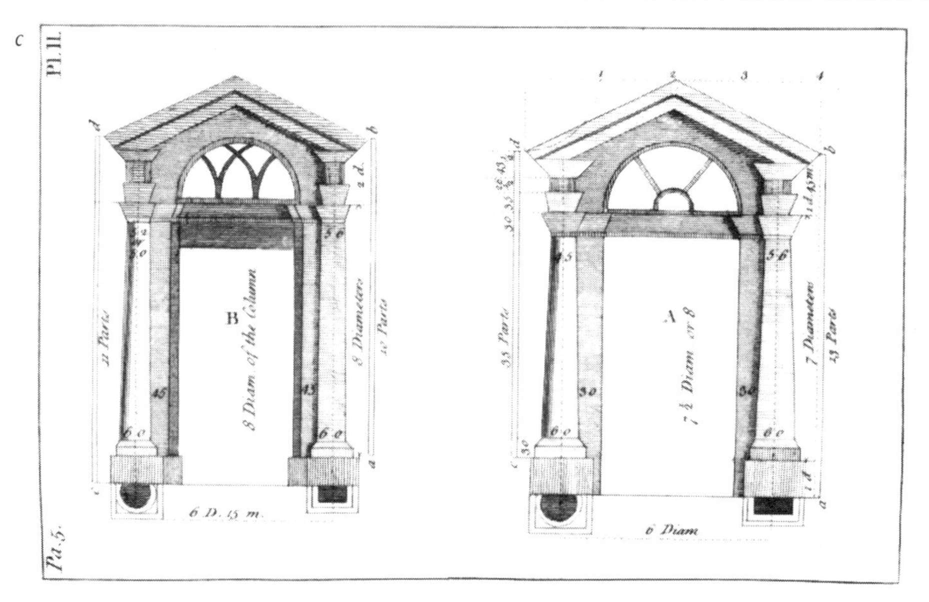

Pl. II.
B
8 Diam. of the Column
8 Diameters
6 D. 15 m.
A
7 1/2 Diam or 8
7 Diameters
6 Diam
Pa. 5.

fanlight, as they usually are in built examples from succeeding decades, and this may imply that no fanlight was intended. Today, doors extending to the full height of a round-headed opening are rare, because they were nearly always reduced later, but one, dated 1725, survives at No. 30 Elder Street, London E1.

In 1757 Robert Morris (cousin to the aforementioned Roger) published an elevation for a terrace house with what may be a metal fanlight and a design for a fanlight with octagonal glazing.[3] Windows with octagonal panes were popular during the 1750s, having been first introduced into the St James's Park elevation of Prime Minister Pelham's house, where the glazing bars were made of brass. The idea was then taken up by Sir Robert Taylor and James Paine, so Morris's designs were quite up-to-date.[4]

John Jores' *A New Book of Iron Work* (1756) and *The Smith's Right Hand* (1765) by W & J Welldon include designs for 'door lights' in wrought iron. How these were intended to be glazed is not made clear and, although multi-paned fanlights in wrought iron were made, some were certainly planted grilles in the Continental manner. Several of Welldon's designs appear to be equally suitable for execution in timber and may have influenced provincial joiners.

The drawings of doorcases in the four Orders in *The Builder's Pocket Treasure*, published by William Pain in 1763, make an interesting comparison with Batty Langley's. Pain's engravings are crude, but they reflect the changes that had taken place in twenty years (p. 8). The fanlight is now integrated into the design in two ways: firstly, it is allowed to break into the pediment, leaving isolated chunks of entablature over the columns, and, secondly, the proportions of the Order are reduced so that the mouldings of the capital continue as imposts for the door arch and as an architrave across the transom. Pain also indicates

Facing page, a. A Doric doorcase from *The Builder's Jewel* (1746) by Batty Langley and, *b.* an executed example, with fretted and spoked fanlight, in Friargate, Derby. *c.* Tuscan and Doric doorcases from *The Builder's Pocket Treasure* (1763) by William Pain (*Victoria & Albert Museum*).

wooden glazing bars in the fanlights, choosing a gothick design for the traditionally rustic Tuscan order and simple radiating bars for the Doric. This type of doorcase may be less classically correct than Langley's, but it sits more comfortably in the façade of the smaller house. There are many examples dating from the 1750s and 1760s in London, and much later in provincial and rural areas – often, to the confusion of historians, added to houses built a good deal earlier.[5] Pain published some further designs in *The Practical Builder* (1774) which reflect the architectural decorations of Robert Adam, but by the mid 1770s the making of windows and fanlights was fast becoming the

Detail from Plate XXXII of *The Practical Builder* (1774) by William Pain (*Victoria & Albert Museum*).

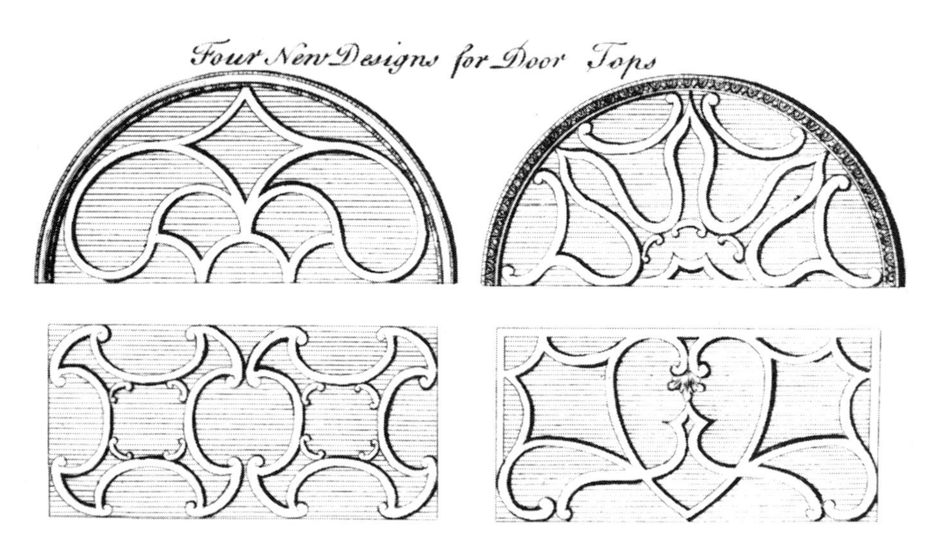

Plate 26 from John Crunden's *Convenient & Ornamental Architecture* (1770). The first design is shown executed on p. 37(c) (*Victoria & Albert Museum*).

province of specialist manufacturers, certainly in London and the larger towns and cities. In fact the publication of exemplars was superseded, in the last years of the century, by a new sort of pattern book in which the author's purpose is not so much to provide designs for craftsmen to follow as to advise potential clients of goods available to order or ready made for sale.

Such is the *Book of Designs*, now in Sir John Soane's Museum in Lincoln's Inn Fields, published *c.*1795 by Joseph Bottomley. The book is prefaced by an 'Advertisement', of which more later, and contains some thirty plates of beautifully engraved designs for fanlights, screen lights – large lights for internal use between hall and staircase – bookcase doors, skylights and all manner of ornamental metalwork (pp. 52–55). Besides Bottomley's catalogue, Soane kept similar publications dating from *c.*1800, by James Cruckshanks, and *c.*1813, by the firm of Underwood, Doyle & Underwood. Also published *c.*1795 was I. & J. Taylor's *Ornamental Iron Work*, a book in the exemplar tradition with several designs for 'Fan Lights' which, in 1795, were unlikely to have been executed. The Taylors were booksellers in High Holborn and Isaac, the father, was probably the same Isaac Taylor who engraved the plates for John Crunden's *Convenient and Ornamental Architecture* in 1770, which included 'Four (Original) Designs for Gothic Door Tops', one of which had been previously published by both Langley and Pain. The most interesting thing about Crunden's work is his use, in 1770, of the term 'door top' when Underwood, in his patent application lodged four years later, was using 'fan light' in a categorical sense (p. 11).

So far we have seen that door lights had to be subdivided by wooden glazing bars, to accommodate the panes of glass available, and that round-headed windows provided an opportunity for decorative treatment. From the evidence of exemplars it would seem that, with the exception of the short-lived fretted type, most fanlights remained fairly simple until the rococo-gothick phase of the 1740s.

Materials: wood

Until *c.*1750 all fanlights, like the windows from which they were derived, were made of wood. There were three types of construction, more than one of which might be employed in a single fanlight. The first makes no distinction between a fanlight and any other contemporary window, except that the glazing bars were usually fitted with the moulded side outwards, so that the less attractive puttied side was seen in silhouette from within (p. 16 a, b).[6] A later refinement, also applied to windows, and which cannot be identified in the field, employs curved members made from laminated timber, that is from a number of thin strips glued together. The second type, exemplified by the fanlight at Marble Hill, involved fretting out the glazed openings from a solid board. Fretting was often combined with glazing bars used like the spokes of a wheel (p. 33). The third type, associated with the rococo-gothick style and characteristic of the 1750s and 1760s, involved hand moulding and carving the glazing bars into fanciful shapes. Fanlights of this type are amongst the most delightful, especially when framed by a fine timber doorcase, and they have the added attraction of being the individual creations of local craftsmen, so that patterns vary from town to town (pp. 35–37). When seen from the inside it will often be found that some of the glazing bars are purely decorative and serve no structural purpose, so that quite elaborate designs may be glazed in only three or four panes. The ultimate development of this trend reduced the fanlight to a decorative wooden grille set in front of the glass. Although technically debased, some attractive designs were made in this way of which a number in Tenterden, Kent, are especially pretty.

In the later eighteenth century hand-moulded wooden fanlights began to be displaced by the products of specialist manufacturers, although wood continued to be used for the simpler designs and in rural areas, where local joiners were not abreast of modern trends. Sometimes a strong local tradition seems to have perpetuated the use of wood. The Georgian New Town of Edinburgh continued to be fanlit in timber when most comparable cities were employing metal, and in Lewes, Sussex, an attractive local variant persisted into the 1820s (p. 34). The change to metal was greatly encouraged, in London, by the Building Act of 1774 which, in order to render the exterior of ordinary houses less combustible (the Great Fire had not been forgotten), banished projecting timber doorcases and required timber frames to be set back within a brick reveal, forming a type of entrance known as a 'door void'. The architects responsible for drafting the Act, Sir Robert Taylor and George Dance, did not, of course, invent this anew, but codified the best practices of the day and ensured that they were followed throughout London. Their insistence on door voids meant that decorative metal fanlights were just about the only permitted adornment, so it is no coincidence that the era of the metal fanlight began in 1775.

Before leaving timber, we must notice one other method of construction in which a wooden fanlight is overlaid and decorated with ornaments made of 'compo', a patent composition consisting, mainly, of whiting and glue. Compo was widely used, from the late eighteenth century onwards, as an inexpensive substitute for woodcarving on architraves, chimney-pieces, mirrors and the like; anywhere, in fact, where it would be hidden under paint or gilding. Robert Adam made much use of compo and established, in 1780, the firm of G. Jackson & Sons, in Hammersmith, who can still supply casts of the ornaments he designed. Because it is not really intended for external use, compo was not often applied to fanlights, but there are some notable exceptions. In Mansfield Street, Westminster, designed by James Adam in the early 1770s, the doorways have sidelights and are treated like Venetian windows to which an outer ring of glazing has been added, so that the fanlight is strongly divided into two rings by a curved timber architrave (p. 42). Every segment of each ring supports a scrolled pendant made of compo, reinforced with iron, and compo is used to enrich the architrave and the central patera. The notion of dividing the fanlight into two concentric rings, as if it were part of a Venetian window, became an important motif during the following decade.

Materials: metal

The building accounts for No. 44 Berkeley Square (better known today as the Clermont Club and for its associations with Lord Lucan) reveal that the windows of the saloon, the grand room on the first floor fronting the Square, were fitted with slim brass glazing bars. This was in 1748 and, since the glass used in the *lower* of each pair of sashes was expensive polished plate, it is clear that the objective was to prevent the glazing bars from impeding the view; this was certainly the case in Henry Pelham's house, which overlooked St James's Park. The point, however, is that it shows just how early in the century experiments had begun into the use of metal glazing bars, albeit for houses of the very rich.

The case for metal glazing bars is set out in the 'Advertisement' prefacing Joseph Bottomley's *Book of Designs*:

> The modern improvements in Architecture are so replete with conveniences, elegance, and taste, that whoever surveys the edifices, erected a little more than half a century back, and compares them with those of the present time, must be astonished at the improvements in this science; amongst the advantages, those of admitting light, are not the least conspicuous. The windows, sky-lights, and fan-lights, of the date before mentioned, are so crowded with wood, as to require a space near double the size of that of the present, to admit the same quantity of light, and air. The change now taking place in the materials for sashes, sky-lights, fan-lights, staircases etc etc from Wood to Metal, has, besides the elegance of appearance, the advantages of strength and extensive durability. The difference in expense in the former and latter, is so inconsiderable, as not to be worthy of notice; nay, in many cases, such as curve-lineal and Gothic work, the expense is less in Metal than in Wood . . .

Bottomley was optimistic in his claim that metal windows cost no more than their timber equivalents,[7] a simple economic fact that ensured the continued use of wood for ordinary sashes, but for 'curvelineal and Gothic work', such as was required in fanlights, metal had obvious advantages.

Most writers have believed that cast iron was the usual metal for fanlights, but this is incorrect. Although cast iron was in use for larger objects, such as railings, it was not until the very end of the century that casting techniques were sufficiently refined, when applied to glazing bars, to offer any advantage in thickness over wood. The earliest metal glazing bars were, in fact, made from brass or wrought iron. Typical sections, not so very different from their timber equivalents and formed, presumably, by a hot-rolling process, are illustrated in William Pain's *The Building Companion and Workman's General Assistant* of 1758. Some fanlights were also made from slim T-sectioned bars of wrought iron, to which cast brass ornaments might be soldered, but the really significant event was the invention of a compound glazing bar in which a thin metal glazing web was soldered to a moulded rib of

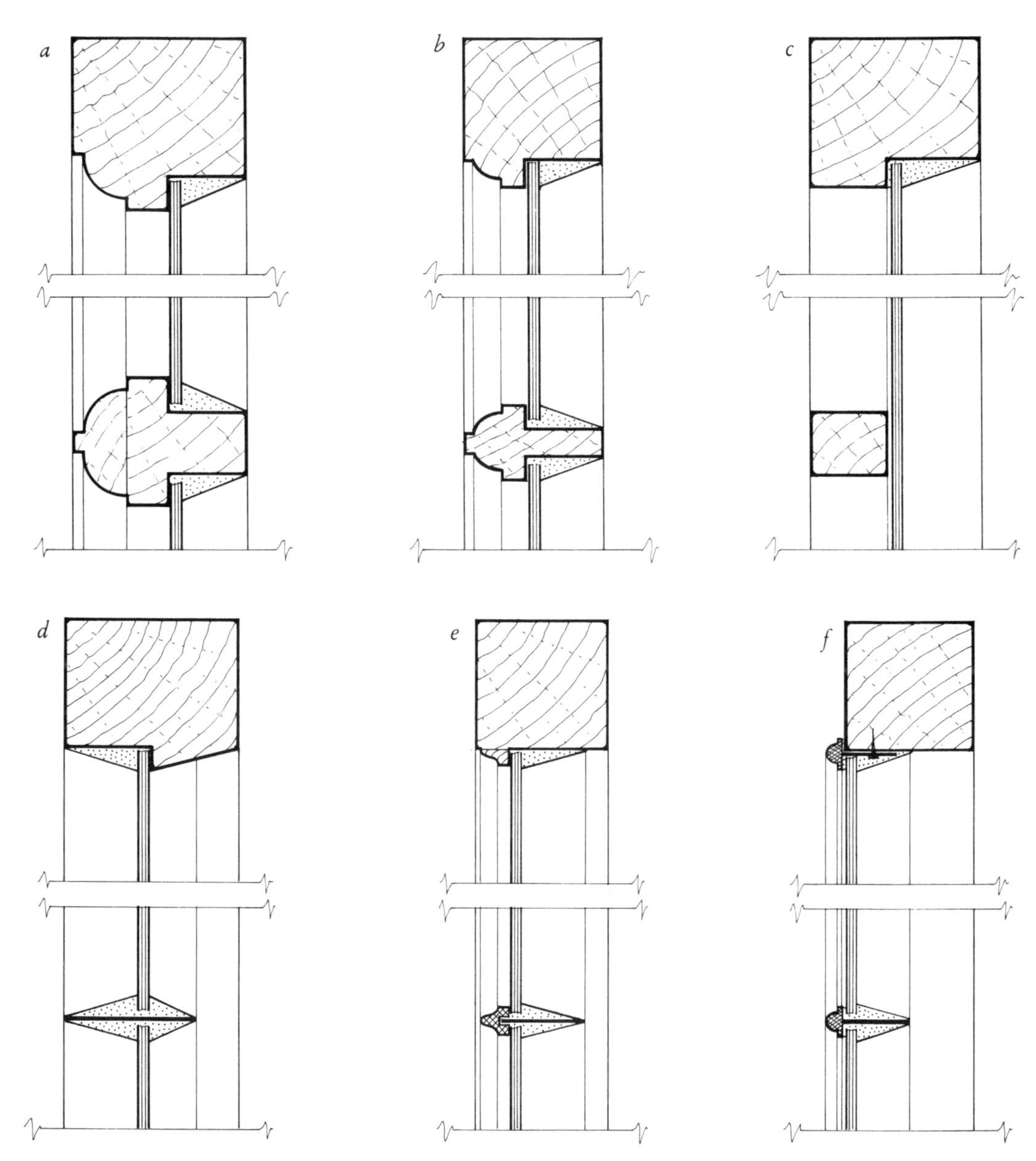

Sections through six types of fanlight: *a*. Early eighteenth century wood; *b*. mid eighteenth century wood, iron or brass; *c*. late eighteenth century fretted wood; *d*. wrought iron and putty; *e*. 'Eldorado'; *f*. Underwood's patent.

lead. The date of this invention is not recorded, but it must have taken place *c.*1770 because several fanlights made in this way survive in a group of houses, in Grafton Street, Westminster, built by (Sir) Robert Taylor between 1768 and 1775. In No. 4 the entrance hall is separated from the staircase compartment by a screen of Doric columns, and Taylor has sufficient confidence, in what can hardly have been an untried form of construction, to place above it a huge semi-elliptical screen-light more than twelve feet wide. A date around 1770 accords quite well with a patent relating to compound glazing bars registered in December 1774 by Francis Underwood, a plumber and glazier from Ampthill in Bedfordshire, since a craftsman would be likely to seek a patent only when his invention had proved successful and threatened to attract unwelcome competitors.

Underwood gives the following description of his process:

> The composition of the said frames is block or grain tin, mixed or melted down with lead to any degree of hardness which the work may require; the barrs or mouldings of the angles for the frames are cast solid in moulds of brass or stone, in some of the three following forms, viz, +, T, L, or any other angles the work may require, and the points of the barrs are afterwards burnt together in a reverse mould. The rabitts of the barrs are either of brass, copper, or tinned iron, or tin plates made in the mouldings before described, and soddered thereto with sodder of the same composition as the mouldings themselves. The points of the mouldings or barrs are let into wood frames, and fastened by the rabbitts with plates and screws or nails, so as not to be seen when glazed and puttied . . .

In other words, the external ribs of the glazing bars were cast in an alloy of tin and lead and soldered to thin glazing webs of brass, copper or tinned iron;[8] where glazing bars crossed or met, the intersecting portions of the ribs were precast to the required angle and then joined to the plain sections by melting the ends together in a stone or brass mould. This construction can be seen in the photograph, on page 18, of a damaged fanlight from Camberwell Grove, in South London. The

Detail of an Underwood fanlight from the back with putty and glass removed. Note rust damage, discontinuous fixing to the sash and unsupported festoon.

photograph also shows the damage caused by rusting of the glazing webs ('rabbitts') and the method of fastening the bars to the sash using thin strips of tin and *tin*-tacks (p. 16 f). In some fanlights examined by the author, the ends of the radiating webs were, as the patent states, actually let into a saw-cut in the sash and the slot filled up with solder.

Seen unglazed this may seem a somewhat flimsy construction, but with the glass in place, preventing the bars from moving sideways, the metal webs impart considerable strength and rigidity to the completed unit – enough, anyway, for large numbers of them to have survived for two hundred years.

Underwood's patent afforded him sole rights in his invention for fourteen years. Early in the 1780s he entered into partnership with Joseph Bottomley and others and, starting with a factory in Padding-

Comparable detail of an 'Eldorado' light showing grooved glazing bars, wooden moulding around the sash and greater use of unsupported ornament. The complete fanlight is shown on p. 57.

ton, the firm of Underwood, Bottomley & Hamble, Fanlight Makers, had moved to substantial premises in High Holborn by 1789 where they remained until 1831. In about 1795, Bottomley left the partnership to set up an independent factory in Cheapside, the original firm continuing as Underwood & Doyle, then Underwood, Doyle & Underwood and latterly as John Doyle & Son. In 1819 Richard Underwood set up on his own in Chiswell Street and, in 1820, Henry Underwood was also making fanlights in Hart Street, Bloomsbury. According to the trade directories[9] there were at least two other specialist fanlight-makers in London – Timothy M'Namara and successors, with premises in Hounsditch, from 1801–30, and James Keir of Soho, of whom more anon. M'Namara probably used Underwood's method, and there may have been others, possibly listed under the more general categories of 'glazier' or 'sash manufacturer', because fanlights of this type continued to be made in great numbers into the 1840s. Outside London there were no large-scale followers of Underwood's patent and the very numerous examples found in the provinces originated in the capital and were, to quote Bottomley, 'executed in good sound workmanship, and of the best materials (glazed if required), and sent to any part of the Kingdom'. This can be proved by a careful study not just of the recurring designs, but of the precise repetition of the applied ornaments. Occasionally an unusual design or ornament may suggest the work of a local glazier or sash-maker but, as a rule, it was cheaper and easier to send to London for anything ornamental. Dublin, on the other hand, must have had its own manufacturers, and some of the designs found in the Irish capital have no parallel on the mainland (p. 47).

If Underwood's invention does indeed date from *c.*1770, he was quick to develop his manufacturing method to the production of large and highly-decorated designs. James Adam's 'Venetian window' fanlights in Mansfield Street, constructed of wood and compo, have already been mentioned, but Robert's beautiful fanlight for No. 20 St James's Square (1771–4) was, according to the building accounts, executed in wrought copper. Underwood must have realised that, by the addition of some cast ornament, his method could be adapted to make such fanlights at a fraction of the cost of wrought work and similar designs in lead appear in Portland Place, a speculative development by the Adam brothers, and in Bedford Square (by Thomas Leverton), both begun in 1776. Portland Place has been sadly mutilated and only a handful of original fanlights survive, including a fine one, stripped of paint and reset in a modern façade, at No. 23. Here the architrave of the notional Venetian window, executed in filigree leadwork, has lost its relationship with the (recent) door below, but in

Bedford Square the decoration on the pilasters dividing doors from sidelights is continued in the fanlights and, in some houses, is further emphasised by being glazed in coloured glass (p. 46).

Besides Underwood's patent there was another successful method of making metal windows invented by James Keir, FRS, sometime around 1780. He established factories at Tipton, near Wolverhampton, and, some years later, in Gerrard Street, Soho. Keir's windows were known as 'Eldorado sashes' after 'a peculiar golden-coloured metal' (actually an alloy of copper, zinc and iron) which he had discovered. In fanlights its use was confined to the cast ornaments,[10] the ribs being made of hot-rolled wrought iron. The glazing bars, like Underwood's, were made in two parts, the webs fitting into grooves in the back of the ribs, where they were retained, seemingly, by crimping the pair together between rollers. The components, including the ornaments, were soldered together and fixed to the sash either with screwed plates or by driving sharpened points into the wood and the border was finished off by pinning a wood or metal half-rib to the inside of the opening in the sash. This treatment of the border provides the easiest way of identifying Eldorado lights because the glazing bars are set slightly below the face of the sash whereas Underwood's stand proud (p. 16 e, f).[11]

Wrought iron and Eldorado metal are stronger than the lead-tin alloy specified by Underwood allowing Keir's fanlights to be glazed in fewer panes, and with more unsupported ornament, so that they can have a delicate, almost spidery, appearance (p. 56).

In 1799 Keir's London factory was taken over by one of his partners, James Cruckshanks, whose catalogue survives in the Soane Museum, and he remained in Gerrard Street until about 1825. Soane regularly specified Eldorado lights in London and they were widely used elsewhere, if not in such numbers as those of Underwood and Bottomley; in Bath, for instance, they are outnumbered by about five to one. An Eldorado fanlight was installed by Robert Adam, in 1786, in Marlborough House, Brighton (Pevsner calls it 'the finest house in Brighton'), and the replacement light on the river front of Marble Hill has already been noted. If, as seems likely, the latter dates from 1781, when Samuel Wyatt made some alterations for Lord Buckingham, it provides a nice link between architect and manufacturer, through Matthew Boulton, FRS, friend and colleague of Wyatt and Keir.

Under Cruckshanks the Eldorado factory did, at last, succeed in marketing a cast-iron fanlight, a simple but ingenious idea for mass-producing lights up to about three-and-a-half feet in diameter. The casting was in the form of a small semi-circle with four radiating arms, which could be cut to accommodate a range of sizes, and it was fixed to the sash with spikes filed from the cut ends. The sashes were made in three pieces, pegged and glued together after assembly round the ironwork; designs were varied by soldering on ornaments and festoons in the usual way (p. 59).

In the 1780s and 1790s metal fanlights enjoyed enormous popularity and the fashion spread rapidly to all parts of the country and overseas to the former colonies in America. Some of the largest and most spectacular examples are to be found in Ireland (see cover) and especially in Dublin, where the door-void-with-Venetian-window type of entrance was adopted at an early date and applied much more extensively than it ever was in England. Indeed, it is quite possible that the fully fledged variety, with columns and entablature, originated in Dublin because some fine examples, complete with spectacular metal fanlights, can be seen in Merrion Square, laid out in 1762. Even if it took a whole decade for the Square to fill with houses, these would still predate James Adam's houses in Mansfield Street, and the fanlights are well ahead of Underwood's patent and his work in Portland Place. The possibility of an Irish origin for compound metal glazing bars is intriguing, and we have already noted the Irish names of Doyle and M'Namara amongst the makers, but the question cannot be resolved until more is known about the dates of certain critical houses.

What strikes the visitor to Dublin, as much as the size and variety of its fans, is the standardisation of its columned doorcases, which persist even in quite modest houses, well into the 1830s, and also the rarity of those rectangular and late patterns which predominate in contemporary London. This conservatism is, presumably, explained by the exodus of fashionable society following the Act of Union in 1801 the very year that Timothy M'Namara first appears in the London Rate Books. In contrast to London, too, some Irish makers favoured flat, unribbed glazing bars and incorporated H-section lead cames into the less structural parts of their designs.

If Dublin has the greater numbers of large fanlights, the most elaborately ornamented were probably those made for two English houses now demolished: Harewood House, London, by Robert Adam, 1776, and Shalford House, near Guildford, 1797 (p. 60). Great fun though it was, Shalford – like some of the Irish fanlights – represents something of a triumph of craft over art; and it makes a nonsense of Bottomley's apologia for metal glazing bars, since it was hardly less 'crowded' with ornament than earlier ones were with wood. Actually the whole of the outer ring was blind, much of it being above ceiling level. Blind, or partially blind, fanlights are not uncommon and many small late Georgian houses have fans that could never admit light. To be properly appreciated they should be painted black with the glazing bars picked out in white, but all too often these days they are glossed over in a single colour.

At the height of their popularity thousands of metal fanlights were inserted into earlier houses. Some designs seem to have been especially attractive for this purpose, and versions of No. 30 in Bottomley's catalogue can be found in Queen Anne and early Georgian houses throughout the country. Space for a fanlight was often created by substituting a six-panel for an eight-panel door, or even cutting off the upper panels. Such alterations can often be detected by looking at the moulding on the inserted transom, because it will be typical of the date of the alteration rather than the doorcase. In one modest house in Wilkes Street, Spitalfields, E1, maximum space for a fanlight was contrived by omitting a transom altogether and cutting a door stop rebate out of the bottom rail of the sash. Nor were fanlights confined to a position over the front door. The exceptional screen light at 4 Grafton Street has already been noted but many large houses had

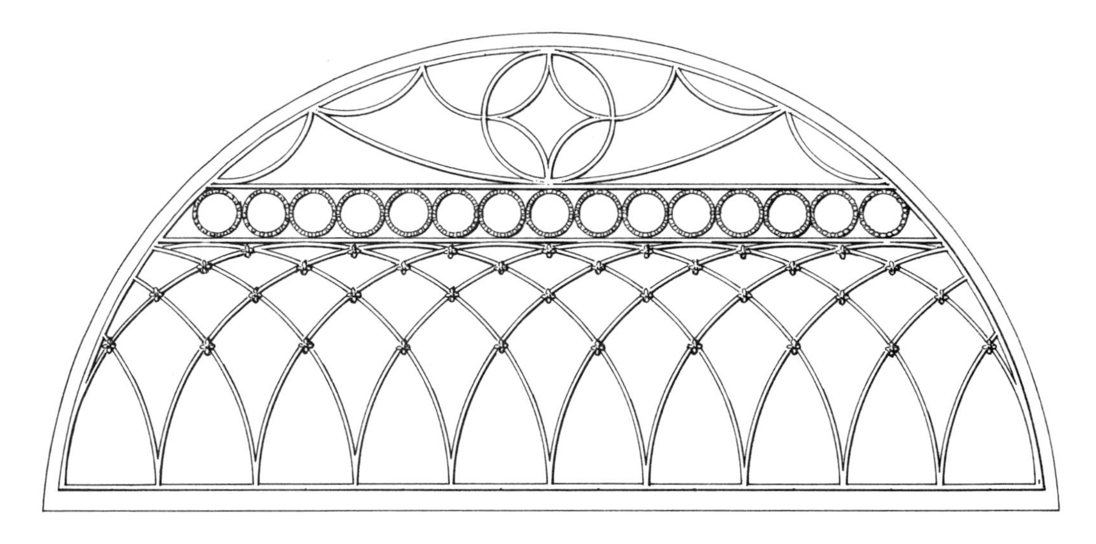

Internal light at 11 Montague Street, London.

secondary glazed doors to exclude draughts from the hall and these were usually surmounted by a fanlight. Commonly this inner fanlight reflected the design of the outer one but, because it was better protected from damage, the opportunity could be taken to add more ornament. There is a stunning example behind the otherwise plain façade of No. 11 Montague Street, just to the east of the British Museum (*above*).

In the eighteenth century metal fanlights were nearly always painted. In the case of Eldorado lights this was necessary to protect the wrought iron from rust, but it was also exceptional for Underwood's products to be left unpainted. Lead may not need protection from the weather but, because it was taking the place of wood, it was painted anyway. A rare exception can be seen at No. 58 Grafton Way,[12] one of a group of houses built *c.*1790 (Nos. 52–6 are replicas). During the Regency period the timber origin of metal glazing bars was, to some extent, forgotten and lead fanlights were sometimes left unpainted as they are in the Nash terraces in Regent's Park.

The late eighteenth century was a time of unbounded experiment, and if the methods of making metal windows pioneered by Francis Underwood and James Keir eventually came to dominate the market, there must have been dozens of others whose methods either failed or achieved only local success. One of the pleasures of fanlight hunting is the constant expectation of finding some new design or unusual construction, such as the fanlight in the house built for himself in 1771 by the Norwich architect Thomas Ivory, which has T-section glazing bars of wrought iron with small scrolls, like miniature Ionic capitals, fixed to the inside. Then there are some fanlights which intrigued the author for a long time because the glass appeared to be supported by nothing but putty; it turns out that the glazing bars are formed from flat strips of iron, their ends jammed into saw-cuts in the frame, and the glazing is indeed held by putty, both inside and out; they must have been built on the bench, first one side then the other, with the panes of glass supported on blocks of wood until the putty had hardened sufficiently to be turned over. Not a sophisticated method, but cheap and, apparently, quite acceptable in rural areas (p. 16 d).

Later designs

During the eighteenth century nearly all fanlight designs were based on a semi-circle and if they were required to fit a rectangular opening it was achieved by filling in the spandrels with extra ornament. Bottomley published a couple of designs based on a square, of which one, No.

umbrella

batswing

30, became very popular. The other had an open ellipse in the centre bearing the street number of a house. If Bottomley's designs are compared with those of Underwood & Doyle, published some twenty years later, the most significant differences are the reduction in the amount of applied ornament and the inclusion of a range of designs based around a circle in which street numbers could be applied in gold leaf. Of these the 'batswing' pattern seems to be the earlier and was, plausibly, derived from an 'umbrella' pattern similar to the one published as a design for a shopfront by John Carter in 1779:[13]

A 'batswing' was used in Doughty Street, London WC1, in the 1790s, within a border of small circles – a motif enlarged to fill a whole fanlight elsewhere in the same street – and again, further north, in Mecklenburgh Square, in 1812, where the sole survivor,[14] No. 17, is also unusual because the central circle is a pivot-hung opening light, or it would be if it weren't so heavily covered in paint. 'Teardrop' fanlights appeared a little later and are the predominant type in the seemingly endless stock-brick terraces of Islington and Camberwell. Even within such apparent uniformity, however, some variety can be observed: coloured glass in the smaller openings or as a ring round the central circle, changes in proportion, different ornaments or, unexpectedly, some novel variation such as the delightful group in Ripplevale Grove, London N1 (p. 75 a).

Underwood & Doyle's catalogue also contains several designs for fanlights to fit rectangular openings of a width exceeding twice their height, a requirement dictated by changing architectural fashions, and designs for fanlights with lanterns. Gas lighting had been installed in parts of London as early as 1792, and it was found convenient to incorporate a small lantern, usually hexagonal in shape, into the centre of the fanlight. These were made and glazed in the same materials except that the domed tops were made of sheet copper or brass, as was the frame of the small glazed door on the inside, giving access to the jet. There are some fine examples, dating from about 1820, in Duncan Terrace, Islington. These handsome 'batswings', with rings of amber glass, were designed and built to take lanterns, but where lanterns were inserted into earlier fanlights they could be contrived rather clumsily, as can be seen in a house on the Old Dover Road in Canterbury (p. 67 b).

Although there had been a move towards simplicity in the earlier part of the nineteenth century, the decorative possibilities of the medium could not be suppressed, and reactions to the ubiquitous 'teardrop' included the introduction of unusual, sometimes ugly, curves and patterns (p. 78). The use of coloured glass became more common, too, and a charming little fanlight, inserted into a house in The Traverse, Bury St Edmunds, has a stained and painted floret leaded into the centre, encircled by segments of mauve, green, amber and blue.

Decline

With the advent of sheet glass in 1832, the structural requirement for glazing bars began to disappear, whilst the quality of crown glass was almost matched by Chance's plate, introduced in 1839. 1840 is the critical date, in London, for the transition from crown to sheet glass, and the change is nicely illustrated in a small area south of the Victoria and Albert Museum. A building agreement for Alexander Square, drawn up in 1827, specifically states that the doorways should have 'large handsome fanlights' and these were duly executed. Pelham Crescent, nearby, was designed by the ill-fated architect George Basevi[15] and built in stages, the eastern half in 1833–8, the western in 1838–43. Basevi's drawings show six-panel doors under semi-circular nine-spoked fans but, in execution, taller, four-panel doors reduced the fanlights, permitting the use of undivided sheets of crown glass. Pelham Crescent suffered some blast damage during the war but one house, No. 13, in the earlier, eastern half, still has a pretty seven-spoked fan, and others may have been destroyed; none of the houses in the western half has an original fanlight, but two retain single panes of early glass. In Thurloe Square, begun in 1840, and Egerton Crescent, in 1843, the overdoors are simple rectangles devoid of glazing bars. Both were designed by Basevi and were, therefore, thoroughly up to date. Ordinary builders were less so as can be seen in the minor streets connecting Thurloe and Alexander Squares: both Alexander Place and South Terrace, built 1841–4, display a variety of metal fanlights, including one that must qualify as the least elegant 'teardrop' in London.

If 1839 signalled the demise of the fashion, it took a long time in dying. Apart from the natural conservatism of speculative builders it seems that some first occupants were either offered the option of fanlights or, perhaps, insisted on them, because they continued to be installed throughout the 1840s. In the country they lasted even longer, and in America they seem never to have entirely gone out of fashion.

The output of the London fanlight-makers during the fifty years to 1840 must have been very considerable, if difficult to calculate. Estimates based on the numbers of deeds registered by the Middlesex Land Registry suggest that annual production for use in new houses in London alone could have reached five thousand in 1790–5 and probably exceeded ten thousand in the boom years of 1824–5.[16] Impressive as these figures are, they take no account of the Georgian expansions of Bath, Bristol, Liverpool, Manchester and a hundred other smaller towns, or of the numerous buildings in the country, which must, together, amount to more than half again. Even these conservative figures yield a grand total in excess of half a million. Assuming that the benefit of modern materials and tools equates, roughly, with the longer hours worked in the eighteenth century, and a fully glazed fanlight now takes three or four days to build, a total workforce of over two hundred must have been engaged in the trade *c.*1825, or perhaps two-thirds of that number if the majority were sold unglazed.

Georgian Revival to the present day

When the Georgian style came to be revived after the First World War the original methods of making fanlights had been forgotten and the concept had become confused with French wrought-iron grilles, introduced into the West End of Edwardian London by architects of the

Beaux Arts (p. 84). Consequently traditional Georgian designs, often with a touch of French ornament about them, were executed in iron or bronze and set up as protective grilles in front of large sheets of plate glass. Of course, in banks and Post Offices, the use of these materials offered advantages in terms of security, whilst the greater scale of many commercial buildings demanded something more robust than the traditional fanlight. There is no denying the beauty and superb workmanship of many of these revival fanlights and they would, perhaps, be better appreciated if they were not seen in an historical context.

In houses most revival fanlights were of the early Georgian wooden type and where metal was used it was less successful because the more intimate scale of domestic architecture requires a delicacy and texture not easy to achieve with flat glass, although the work of one London maker, using wood and compo glued to sheets of plate glass, requires close inspection to reveal the trick. Yet correctly detailed lights could be made, as is shown by Lloyd's Bank in Rye (domestic in scale if not in purpose), built in 1920, where the beautifully-made replica over the staff entrance is perfect in every detail and is only given away by a certain quality in the carving of the patterns.

It is a great pity that similar standards have not been applied to so much of the restoration done in recent years; most of the new fanlights installed since the war have either been metal grilles installed in front of, or, even worse, behind, plain glass, or made up as leaded lights (p. 85). How some of these came to be installed in buildings officially listed as of architectural interest defies understanding since they bear little resemblance to the originals and do nothing to enhance the appearance of the buildings. A surprising number of large, and rather well executed, leaded-light fans have been installed in and around Harley Street, where they almost constitute a local species, but it is to be regretted that the same effort and craftsmanship have not been devoted to the production of authentic replicas.

No review of this subject would be complete without some reference to the proliferation, during the last ten years, of hardwood doors containing a semi-circular fanlight. Leaving aside ecological issues, involving the frivolous use of irreplaceable foreign hardwoods, such doors are not acceptable as replacements for traditional softwood doors in older houses. Apart from untreated English oak, doors and windows in this country have always been protected from the weather by paint, so sealed hardwood, which (despite the manufacturers' claims) quickly deteriorates, strikes an incongruous note in a street of painted doors. More than a coat of paint, however, would be needed to disguise the unsuitability of these designs; if the original door opening is rectangular, then introducing a semi-circle into it is out of place, and if it already has a fanlight, decorative or plain, placing another one below it is ridiculous. Poor taste is not, of course, confined to the twentieth century, for the same arguments can be levelled at some of the eighteenth-century inserted fans (p. 65). There are, too, Georgian precedents for doors with built-in fanlights, including one in a mediæval doorway in The Close at Norwich, and others in Bath and Poole. A country house near Falmouth has a fanlight on top of the door but, as the opening is round-headed, this is merely a way of saving the space occupied by a transom.

Survival

Of the enormous number of fanlights made between 1780 and 1840, relatively few have survived. Some have succumbed through age, poor workmanship and the weather; sea air seems to affect them badly, for otherwise it is difficult to explain their scarcity in such popular Regency resorts as Brighton, Broadstairs and Ramsgate. Bombing during the Second World War must have accounted for many hundreds, even where the buildings themselves remained standing, but most have perished through redevelopment, neglect and, in town centres, the replacement of ground floors by shops. The latter is especially noticeable in towns which are large enough to attract the leading retailers without the space to accommodate them with ease. Ashford, in Kent, is a town where hardly a fanlight survives despite a High Street composed, above ground-floor level, of many fine Georgian buildings.

And the case was made worse when most of the housing around the centre was swept away for a ring road, car parks and modern office buildings. Nor has time dealt kindly with Dublin fanlights and, of those that remain, a significant number are distorted in a way that suggests settlement of the piers supporting the door arch; built in a climate of social competition, more attention was perhaps paid to a showy fanlight than to the more prosaic considerations of providing adequate foundations.

Changes in fashion may have led to the actual removal of fanlights in favour of plain glass, a lantern or even a stained glass overdoor, but more significant are the factors which have made whole areas of our larger cities unfashionable. Fine houses slowly descending the social scale, suffering the gradual attrition of their original features, are a sad reminder of other, if not necessarily better, days. Parts of London, Dublin and much of Liverpool impart this nostalgic feeling of decayed gentility – until, suddenly, one comes upon Liverpool's Rodney Street and is given a glimpse of the elegance of the Georgian city. The fortunes of Liverpool have declined with those of its port, while much of London has been lost either to bombing or post-war clearance. The same excuses cannot be made for Dublin, where, despite the activities of the Irish Georgian Society, the almost wilful destruction of the most complete Georgian city proceeds apace. Thankfully the Georgian New Town of Edinburgh and the City of Bath now appear to be safe, but neither of them ever exhibited the variety, or sheer size, of fanlights found in Dublin and London.

a

b

EARLY FRETTED FANLIGHTS: *a*. Marble Hill House, Twickenham, 1725; *b*. Bedford Row, London, *c*.1720.

SPOKED FANLIGHTS: *c*. Castle Street, Farnham; *d*. The Close, Norwich.

c

d

a *b*

LOCAL STYLE IN TIMBER Early nineteenth-century doorways in Lewes *a*. High Street *b*. West Street. Late eighteenth-century fretted lights typical of *c*. Bury St. Edmunds (Guildhall Street); *d*. Farnham (Castle Street).

c

d

EAST ANGLIAN FRETS: The Close, Norwich, *a*, and Bury St. Edmunds; *b*. Westgate; *c*. South Hill; *d*. Sparhawk Street and, *e* & *f*. St. Mary's Square.

FRETS FROM THE SOUTH-EAST: *a.* Montpelier Row, Twickenham; *b.* (dated 1774) & *c.* High Street, Sandwich; *d.* & *e.* High Street and, *f.* Ashford Road, Tenterden.

a

b

a. Delicate tracery in unidentified material in Ashford Road, Tenterden; *b*. Wood bars and metal scrolls in James Street, Macclesfield.

a

b

Wood with 'compo' beads in *a*. Great Pulteney Street, Bath, and *b*. Park Street, Macclesfield.

a

b

c

d

EARLY WROUGHT IRON: *a, b.* Mid eighteenth-century examples in Bedford Row, London. Note unribbed glazing bars, missing ornaments and rust damage.

1770S METAL: *c.* Wrought iron with scroll capitals at Ivory House, Norwich, 1771; *d.* copper at 20 St James's Square, London, *c.*1771–4.

a

'VENETIAN WINDOW' DOORCASES: *a*. 15 Mansfield Street, London, executed in wood with 'compo' ornament, and, *b*. a similar design in metal in Harcourt Street, Dublin. This doorway, $9^{1}/_{2}$ feet in diameter, shows evidence of the structural settlement not uncommon in Dublin.

a

b

a. Dublin doorcases, like this one in Merrion Square, have entire columns and entablature; *b*. in London these elements tend to be stylized, as in Bedford Square.

a

b

a. A replica fanlight, by the author, in Portland Place; *b*. detail from Bedford Square showing some ornaments copied in the replica.

a

b

a. Several ornaments in this example in Gardiner Row, Dublin, are shared with Bedford Square whereas the motifs in *b*. in Merrion Square are peculiar to Ireland.

a

b

a, b. Two late eighteenth-century fans from Gloucester Place, London. *c*. A 'Dublin' doorcase in Rodney Street, Liverpool.

c

a

b

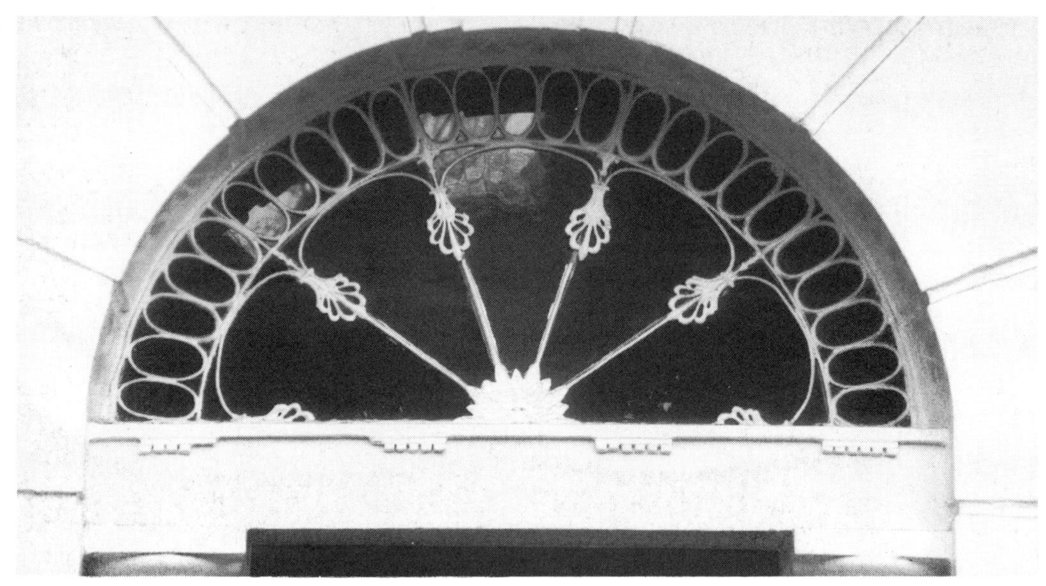

a. Simple fan in Wimpole Street; *b*. decorated example in Montagu Square, London.

VARIOUS FANS: *a*. West Street, Farnham; *b*. Kennington Park Road, London; *c*. Marlborough Buildings, Bath; *d*. Rodney Street, *e*. Upper Paul Street & *f*. Falkner Street, Liverpool.

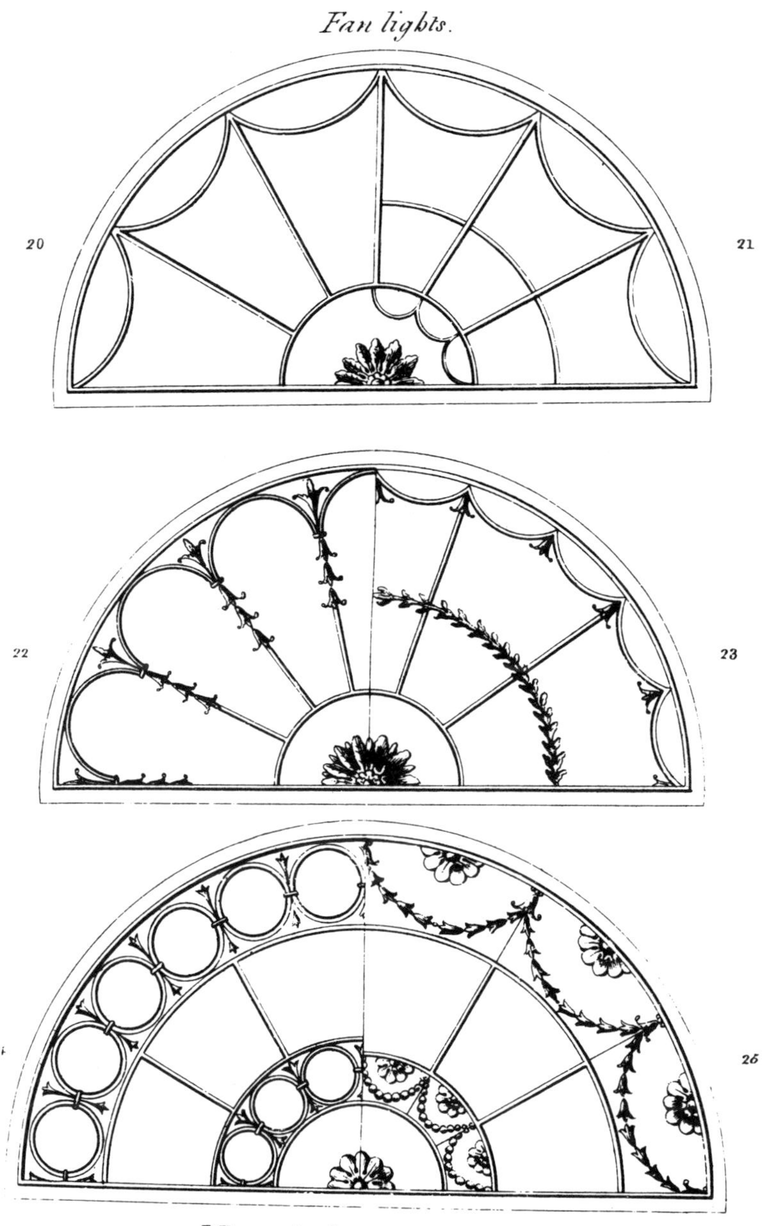

J. Bottomley Inv.^t & del.^t Decr 1793.

30

31

(*facing page*) Designs 20–25 from Joseph Bottomley's catalogue of *c.*1795 and, (*above*) designs 30 & 31 (*Soane Museum*).

a

b

a. Bottomley's designs 28 & 29 (*Soane Museum*); *b*. design 29 as executed in High Street, Newmarket.

a

b

a. Bottomley's design No 8 (*Soane Museum*); *b*. close copy in Montagu Square, London.

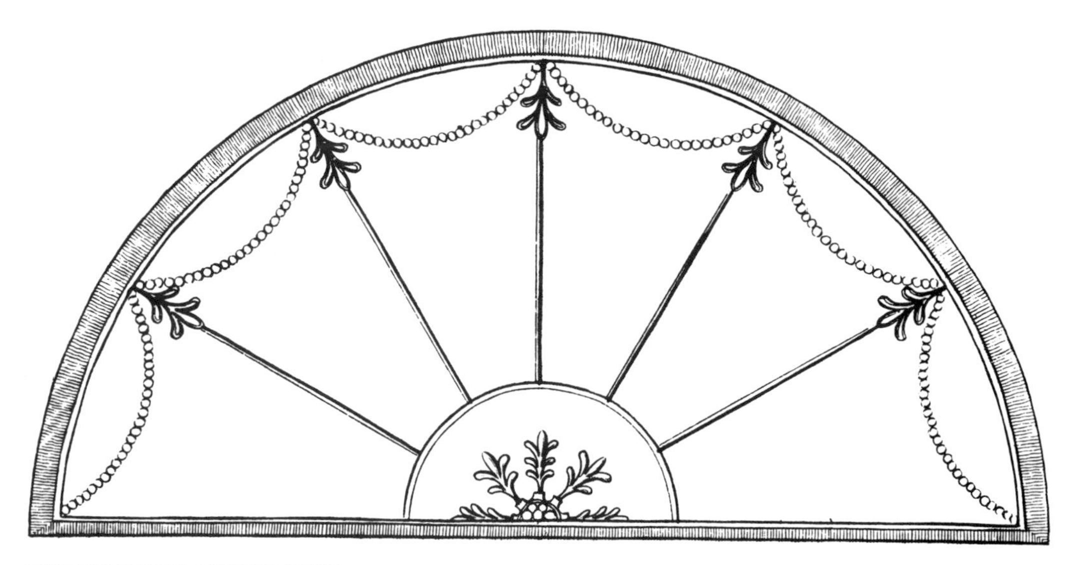

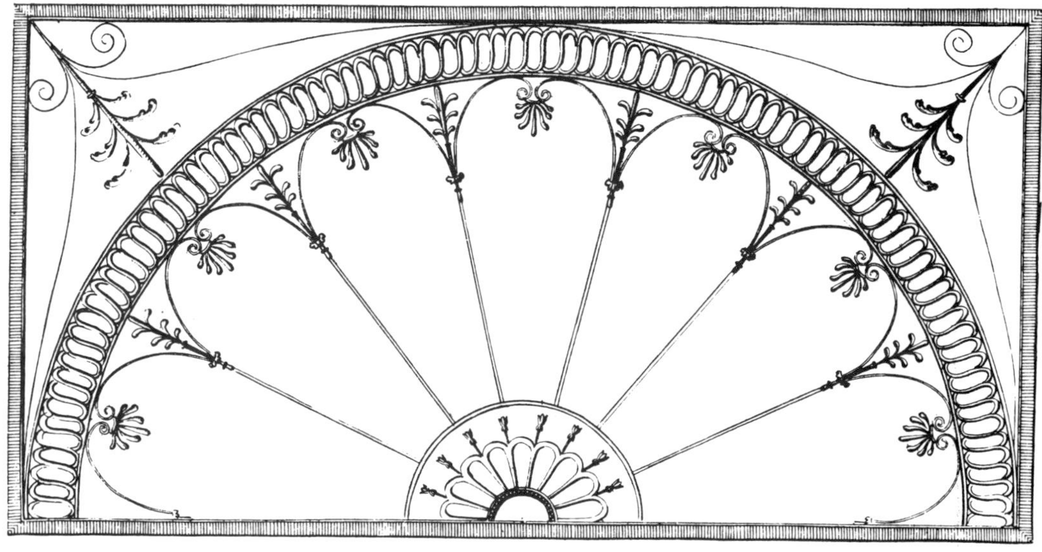

Designs 3 (*top*) & 12 from James Cruckshanks' *c.*1800 catalogue of 'Eldorado' lights (*Soane Museum*).

a

b

a. An 'Eldorado' fanlight in Bridge Street, Worcester; *b*. the same during restoration. Note the use of ornaments from design 12.

ELDORADOS IN BATH: *a*. Spidery example in Park Street; *b*. 'Blind' light in Laura Place; compare central floret with design No 12 on p. 56.

'Eldorado' fanlights of different diameters made from a common casting: *a*. Palace Street, Canterbury (the central floret is damaged, see design No 3 on p. 56); *b*. Warren Street, London (central floret replaced by author).

a

b

ENGLISH ORNATE: *a*. Harewood House, London, by Robert Adam, 1776 (*Victoria & Albert Museum*); *b*. Shalford House, Guildford, 1797 (*National Monuments Record*).

IRISH FLAMBOYANT: *a*. Hume Street, Dublin; *b, c.* Leeson Street, Dublin.

a

b

1790S RECTANGLES. *a*. Laura Place, Bath: this design, adapted from a Bottomley sidelight, also occurs in Farnham and Brighton; *b*. Castle Hill, Dover.

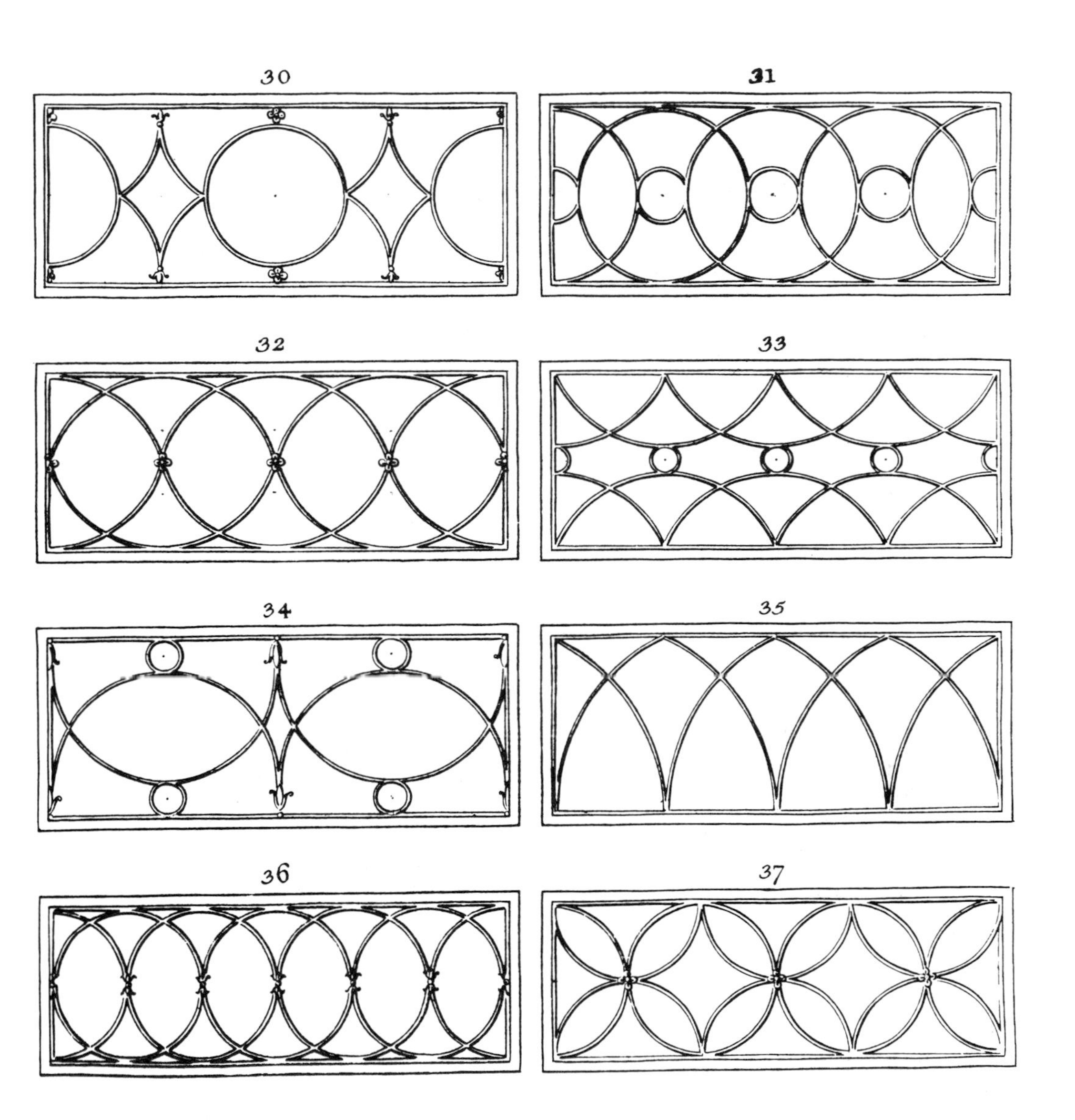

Designs for rectangles from Underwood & Doyle's catalogue, *c.*1813 (*Soane Museum*).

(*Soane Museum*)

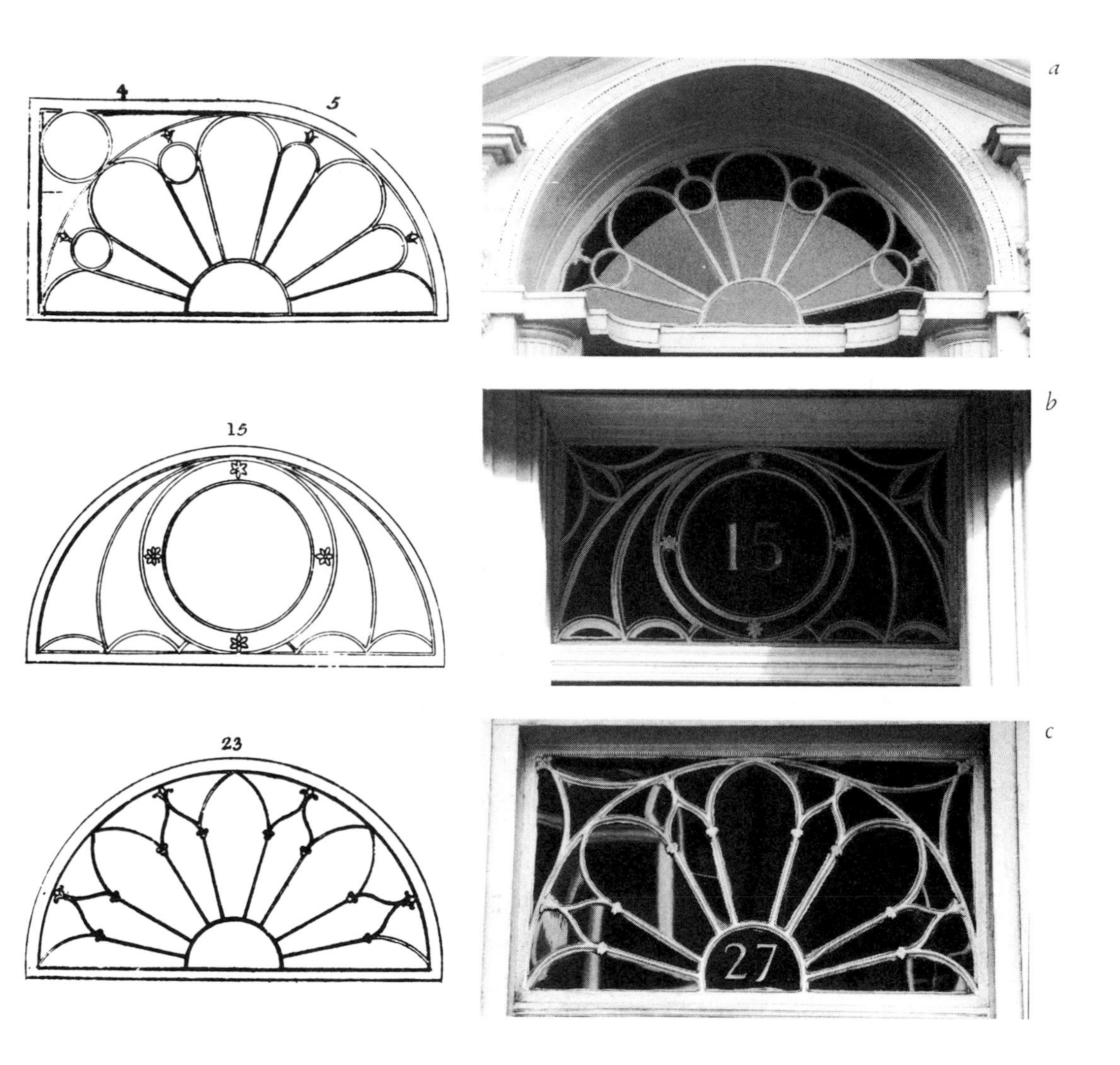

Three designs from Underwood & Doyle's catalogue of *c.*1813 (see *facing page*) with executed examples. *a.* Castle Street, Farnham; *b, c.* Great James Street, London. Note the way a semicircular design is adapted to fit a rectangular opening.

a

b

c

INSERTED LANTERNS: *a*. Typical Dublin lantern in Merrion Square; *b*. Old Dover Road, Canterbury. *c*. Portland Place, London.

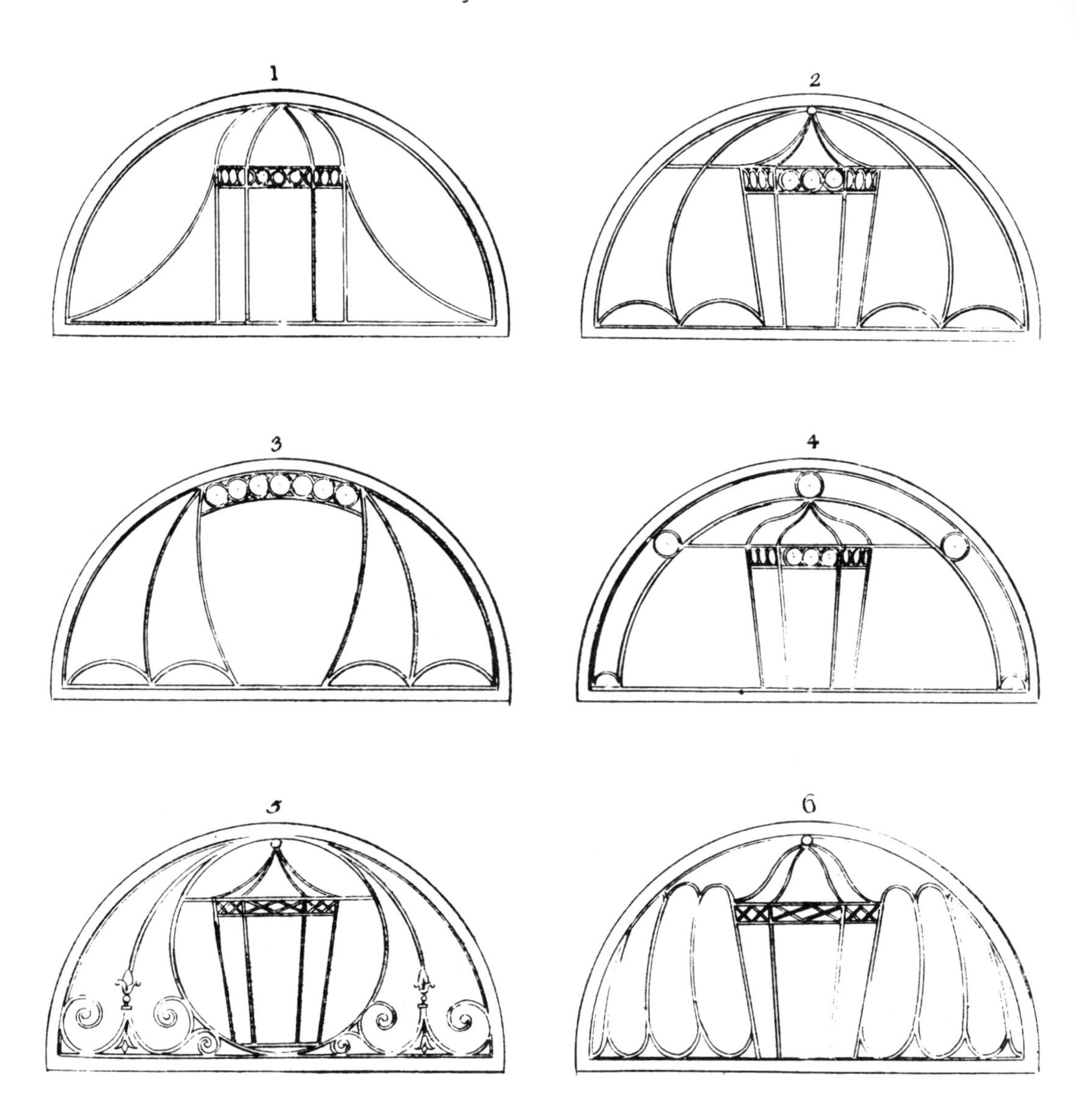

Fanlights with lanterns from Underwood & Doyle (*Soane Museum*)

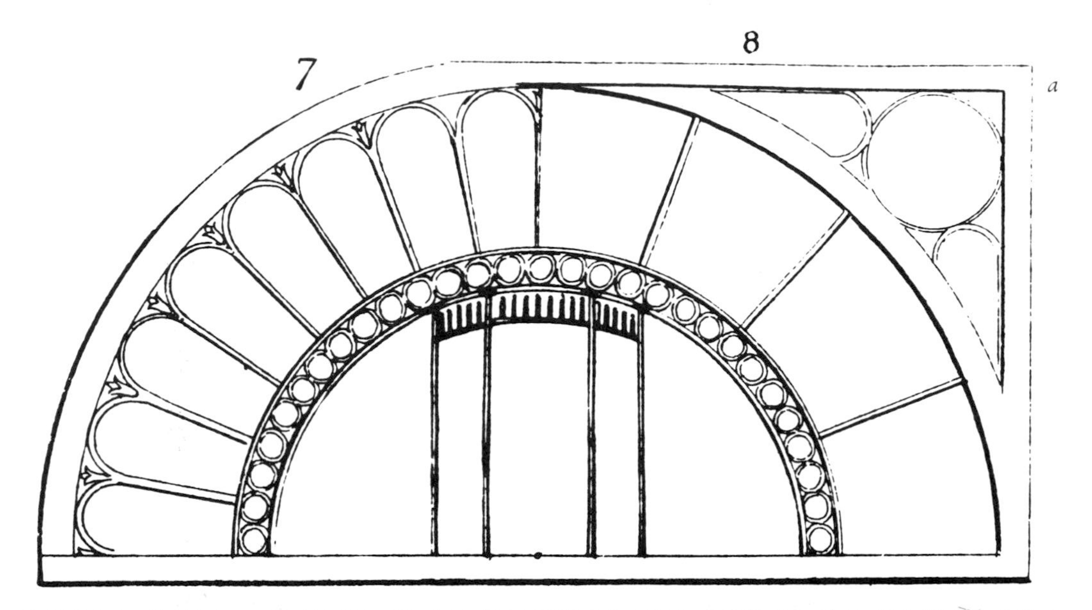

a. Designs No 7 & 8 (*Soane Museum*); *b*. Close copy of No 8 in Cadogan Place, London.

a

b

DUBLIN LANTERNS: *a*. Pembroke House, Pembroke Street; *b*. similar lantern in a wooden fanlight in Merrion Square.

a

b

ENGLISH REGENCY: *a*. The Paragon, Bath; *b*. Devonshire Place, London.

a

b

c

REGENCY 'BATSWING': *a*. Upper Wimpole Street, London. The circles in the sidelights continue as a decreasing border round the fanlight. *b*. 'Batswing' in Gloucester Place, London; *c*. Stilted 'teardrop' in Amwell Street, London.

FOUR 'TEARDROPS': *a*. Marine Square, Brighton; *b*. Pembroke Square, Kensington; *c*. Ripplevale Grove, Islington; *d*. Camberwell New Road (b & d are author's replicas).

a. 'Teardrop' variant typical of Ripplevale Grove, Islington; *b*. reversed 'teardrop', with opening light, in Alfred Street, Bath.

a–d. Reversed ‘teardrops’ in Alexander Place, Kensington, Camelford Street, Brighton, Duke Street, Norwich and Friargate, Derby; *e, f*. ‘batswing’ variants in Severn Terrace, Worcester and Gibson Square, Islington.

SIX LATE DESIGNS: *a*. Oak Place (1836), Tenterden; *b*. Bewdley Street, Islington; *c*. Glebe Place, Chelsea; *d*. Bridge Street, Pershore; *e*. Rodney Street, Liverpool; *f*. Chequer Square, Bury St. Edmunds.

SIX UNUSUAL DESIGNS: *a*. Hume Street, Dublin; *b*. Whiting Street, Bury St. Edmunds; *c*. London Road, Dover; *d, e*. Annett's Crescent and Duncan Terrace, Islington; *f*. The Close, Norwich.

a

b

a. A singular design in Fitzroy Square, London; *b*. delicate pattern in Sterling Street, Knightsbridge, repeated as an oval window in The White House, Whitton, Suffolk.

a

b

FANLIGHTS AS WINDOWS: *a*, *b*. Bowed shopfronts in Argyle Street, Bath. The lead fanlights over the windows in *b* show that they should be glazed in 5 × 3 panes; the side door has an 'Eldorado' fanlight.

b

LONDON AND DUBLIN COMPARED: Terraces of similar date in *a*. Mount Street Crescent, Dublin, and *b*. Camberwell New Road, London, showing the persistence of fans and entire columns in the Irish capital.

a

b

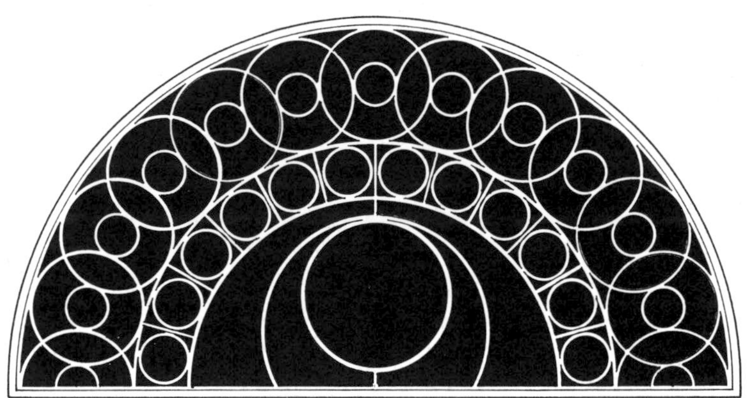

A 'blind' fanlight in Bridge Street, Pershore. *a*. As it was in 1988; *b*. as it should look with the glazing bars picked out in white.

NEO GEORGIAN: Bronze fanlight in a 1920s bank in Sandwich (*Neil Burton*).

a

b

a. Wrought iron grille in Portland Place, London; *b*. Lead-light replica in the same street; note that most of the ornaments have fallen off.

a. Even a well-made lead-light fails to convince because the bars are unribbed, it lacks depth and, contrary to period practice, it is unpainted. *b*. This replica employs lead strips cemented to a single sheet of glass and leaves the central circle without apparent support.

a *b* *c*

FANLIGHTS IN DOORS: *a, b.* Georgian precedents from Church Close, Poole and The Close, Norwich; *c.* the ultimate absurdity.

NOTES

1. Over the front entrance; that on the river front was replaced *c.*1781 by an 'Eldorado' fanlight.
2. Batty Langley, *Ancient Architecture Restored and Improved ...*, 1741.
3. W. Halfpenny, Morris & Lightoler, *The Modern Builder's Assistant*, 1757.
4. Many octagonal sashes have been replaced but their former existence can be deduced from octagonal panelling on the shutters, as in the Cabinet at Felbrigg, Norfolk, altered by Paine in 1751.
5. Pain was still producing new drawings of this sort of doorcase as late as 1789, in his *Practical House Carpenter*, fifteen years after projecting timber doorcases had been prohibited in London by the Building Act of 1774.
6. This practice was also followed in shop windows which, likewise, presented their fairest face to the street.
7. H. J. Louw has extracted figures from Soane's Bill Books which indicate that mahogany sashes (with frames) cost 2s 10d per square foot compared with 3s 4d for Eldorado sashes.
8. In practice tinned iron was preferred because it was cheaper.
9. There were several early London directories; each listed subscribers by name, making a company that changed hands difficult to follow.
10. In later years ornaments of various materials were used including pressed iron and lead-tin alloys. Some ornaments were common to Eldorado and Underwood/Bottomley lights suggesting that the patterns were generally available, perhaps as 'compo' casts obtained from Jacksons.
11. The author is guilty of confusing this distinction by making replicas of Eldorado lights with lead ribs conforming to Underwood's profile.
12. Strictly this light belonged to No. 52 and was moved to its present site in 1985.
13. John Carter, *The Builder's Magazine*, 1779, Plate LXIX.
14. Nos 44–7 are replicas made before research had established that the border of No. 17 was incomplete. It should be decorated with Greek fret.
15. He was Surveyor to Ely Cathedral and fell to his death from the Lantern in 1845; it is said that he might have saved himself but for having his hands in his pockets.
16. Francis Sheppard, Victor Belcher & Philip Cottrell, 'The Middlesex and Yorkshire deeds registries ...', *The London Journal* Vol.5 No.2, 1978.

BIBLIOGRAPHY

Books

Clifton-Taylor, Alec, *The Pattern of English Building* (London, 1965).
Six More English Towns (London, 1981).

Cruikshank, Dan and Peter Wyld, *London: The Art of Georgian Building* (London, 1975).

Draper, Marie P. G., *Marble Hill House and its Owners* (London, 1969).

Harris, John, *English Decorative Ironwork from Contemporary Source Books, 1610–1836* (London, 1960).

Kearns, Kevin Corrigan, *Georgian Dublin* (Newton Abbot, 1983).

McGrath, Raymond and A. C. Frost, *Glass in Architecture and Decoration* (London, 1937).

Summerson, John, *Georgian London* (London, 1962).

Survey of London passim.

Swarbrick, John, *Robert Adam and his Brothers* (London, 1916).

Articles

Louw, H. J., 'The Origin of the Sash Window', *Architectural History* 26, 1983, pp. 49–72.
'The Rise of the Metal Window', *Construction History* 3, 1987, pp. 31–54.